BRITAIN'S HERITAGE COAST

The JURASSIC COAST

PAUL HARRIS

AMBERLEY

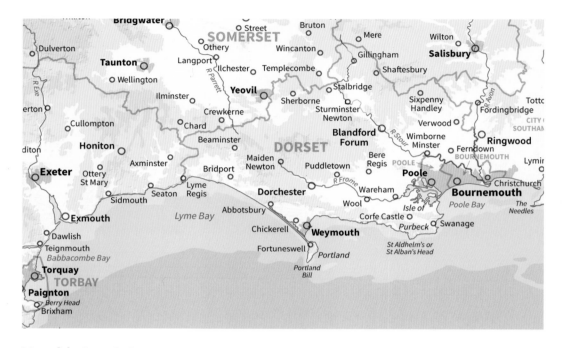

Map of the Jurassic Coast.

First published 2014

Amberley Publishing
The Hill, Stroud, Gloucestershire, GL5 4EP
www.amberley-books.com

Copyright © Paul Harris, 2014

The right of Paul Harris to be identified as the Author
of this work has been asserted in accordance with the
Copyrights, Designs and Patents Act 1988.

ISBN 978 1 4456 1917 0 (print)
ISBN 978 1 4456 1922 4 (ebook)

British Library Cataloguing in Publication Data.
A catalogue record for this book is available from the
British Library.

Typesetting by Amberley Publishing.
Printed in Great Britain.

Contents

Acknowledgements

I would like to thank all those who have provided information, suggestions, photographs, advice, encouragement or practical assistance that has helped in the preparation of this book. My particular thanks go to George A. Wilkins of the Norman Lockyer Observatory in Sidmouth for providing me with the full story relating to Donald Barber, which many of you will learn about for the first time in Chapter 1; to Graham Davies of the Lyme Regis Museum and Mike Applegate for permission to use the picture of the cottage uplifted by the 1840 Whitlands landslip, also in Chapter 1; to my partner Candida for the many photographs she took and contributed to this book, for her happy company on my later visits to the Dorset coast, for her encouragement when my energy or enthusiasm flagged and for finding out useful information I would not even have considered looking for; to Brenda Gokreme for the loan of her ancient copy of *The Geology of Weymouth and the Isle of Portland (with notes on the Natural History of the Coast and Neighbourhood)*; to Ray Hollands for the use of his photographs in Chapter 3; to my sister, Maureen Barkworth, and brother-in-law, Brian Barkworth, for the photographs and information they contributed; to Natalie Earl and The Landmark Trust for supplying pictures and information regarding Clavell Tower in Chapter 5; to Mike Smith of the National Coastwatch Institution (NCI) and Rupert Wood of the Maritime Volunteer Service (Poole Unit) for the picture of one of their exercises at St Alban's Head and other relevant information relating to NCI activities kindly supplied in Chapter 5; to Andrew Wright of Swanage Railway Trust, the Swanage Tourist Information Centre, Robin Boultwood, Becky Stares and Jackie Lane for the pictures kindly provided for use in Chapter 6; to Christine Heald for her company on some of my earlier visits to the Jurassic Coast; to my good friend John Webber for helping with the practical support that ensured this project was completed on time; and of course to Amberley Publishing for commissioning me to write this book in the first place and producing the copy you see before you.

All pictures not otherwise credited were taken by the author. I have naturally tried to make sure that all sources have been credited correctly, but if there are any mistakes or omissions then please contact the publishers to ensure the necessary revisions are included in future editions of this work.

Introduction

Many years ago when I was just twenty, inspired by scenes in the 1967 film version of Thomas Hardy's *Far from the Madding Crowd,* I conceived the idea of walking the Devon and Dorset coast from Sidmouth to Swanage. I tried it over one long weekend, which, of course, was far too ambitious and allowed little time to look around and explore fully. I got as far as Weymouth before running out of time, but I was determined to come back and continue the journey when time and circumstances allowed. That I did, on numerous occasions over the following forty years, each time exploring short sections of this coast in more detail.

In 2001, the spectacular 95 miles of coastline between Exmouth in Devon and Studland in Dorset, now known as the Jurassic Coast, were designated a UNESCO World Heritage Site – the first British natural landscape feature to be granted this status. The reason for this was basically down to the geology and geomorphology of the area. The cliffs along this coast expose 185 million years of the earth's history in their strata, covering three geological periods; the Triassic, Jurassic and Cretaceous. Each period's sedimentary deposits reveal an unparalleled fossil record of that time and its marine and terrestrial life. This record is constantly coming to light through coastal erosion, drawing in fossil hunters from across the world. The wide range of rock types and constant erosion have also created an extremely varied coastal landscape from a scenic as well as a scientific point of view. Consequently, this landscape has importance as a feature of outstanding interest. But it's not all about geology and scenery.

The Jurassic Coast exhibits a wide range of habitats from a natural history perspective. From the vegetated undercliffs of Hooken Cliffs, the Axmouth to Lyme Regis undercliffs and White Nothe at Ringstead Bay, to the sheltered waters of the Fleet behind Chesil Beach, the harsh stone landscape of the Isle of Portland and the rich marine environment that exists along the length of this coast, there is much to draw the interest of the naturalist.

There is a human story too: artists, writers and sculptors, among them, Michael Fairfax, Sir John Everett Millais, Jane Austen, John Fowles, John Meade-Falkner, Ian McEwan, John Piper, Sir Antony Gormley, Ian Fleming, Paul Nash, Enid Blyton and, of course, Thomas Hardy, have all found inspiration here. Generations of quarrymen have mined this coast, prisoners have helped build Portland Harbour, troops practised for the D-Day landings and Olympic and Paralympic participants competed in Portland Harbour and Weymouth Bay, while members of the National Coastwatch Institution have kept an eye out for all those who frequent this coastline and its waters.

With so much that can be said about the Jurassic Coast, it is not surprising that over the years, and especially since 2001, there have been a number of walking guides, maps, leaflets, websites and general guidebooks devoted to this area, all full of brief facts, directions, useful contact numbers and beautiful photographs.

An excellent range of leaflets and books are produced by the Jurassic Coast Project. There seems little point in merely replicating the sort of thing that has already been done more than adequately elsewhere. Also, as someone who is not only a keen walker but also enjoys armchair travelling, I find most of these, though useful, somewhat unsatisfying. All too often, they concentrate on the essential information required by the visitor, including little more than a brief mention of the many interesting or unusual features, sights, stories, natural phenomena, history and literary connections one can encounter or learn about along the way. This book, then, is conceived as an attempt to address that need by delving into some of the lesser known (as well as the well-known) features and aspects of this very special coastline's natural and human history, its cultural connections and claims to fame. The choice of subjects covered is a personal one, as is the direction of travel.

This book has been arranged as a geographical journey from west to east, following the shoreline (occasionally with slight deviations) and keeping the sea to my right. Why keep the sea to my right, you may ask, and not travel in the other direction? That can only be answered by saying that this is my personal preference. I always tend to walk coastlines with the sea to my right. This might be something to do with having the prevailing wind at my back or walking towards the rising morning sun (at least on the broadly south- or east-facing coasts). It may be connected to eyesight, balance, being right-handed or some other subliminal psychological or physical reason I have yet to recognise. Of course, you the reader may prefer to travel in the other direction, moving east to west with the sea to your left – take your pick. Hopefully my book will still be of interest to you.

However you decide to travel along the Jurassic Coast, my hope is that this book provides an inspiring flavour of this varied coastline, its nature, history and atmosphere, and that this helps make exploration by both the appreciative visitor and local resident alike more rewarding.

Paul Harris, Folkestone

1

Orcombe Point
to the Undercliffs

The 95-mile stretch of shoreline known as the Jurassic Coast, by virtue of its exhibiting 185 million years of geological history within the strata of varied and spectacular cliffs, starts (if heading eastwards) just outside Exmouth in Devon at the headland known as Orcombe Point. The precise spot is marked by a striking 'geoneedle' sculpture made from examples of the different rocks to be found along the Jurassic Coast. It was created by Michael Fairfax and unveiled by HRH The Prince of Wales in 2002. This was the year after the Jurassic Coast was awarded UNESCO World Heritage Site status, Britain's first natural landscape feature to be recognised in this way.

Despite being named after the most well-known of the three geological time periods of the Mesozoic era, this coastline comprises of landscape formed across them all. The earliest of these periods, the Triassic, commenced around 250 million years ago. This was followed by the Jurassic and finally the Cretaceous period, which ended abruptly around 64 million years ago, possibly as a result of a major asteroid impact in what is now the Gulf of Mexico.

The cliffs east of Orcombe Point owe their red colour to Triassic sandstone, and are said to be the richest site of Triassic reptile remains in Britain. The coast from here to Lyme Regis is also known as the East Devon Heritage Coast. The crumbly nature of this coastline is well illustrated by the sliding cliffs known as The Floors, east of the large caravan site at Sandy Bay. Here, a coastal landslip forms level shelves down the cliff face, a feature that will be encountered many times on this journey by virtue of the creation of undercliffs, areas of level, vegetated land between the cliff edge and the sea.

Budleigh Salterton and its Most Famous Son

The first town on our journey is the genteel Victorian resort of Budleigh Salterton. This now sedate little town had a busy harbour until this silted up in the fifteenth century. The town's most famous son was undoubtedly Sir Walter Raleigh, who was born in 1554 2 miles north of the current town at a farmhouse at Hays Barton, East Budleigh. This soldier, courtier and explorer, a favourite at the court of Queen Elizabeth I, was knighted in 1584. He led numerous expeditions to the Americas between 1584 and 1595, leading to the establishment of the colony of Virginia (named after Elizabeth I, the 'virgin queen') and to the introduction of both tobacco and potatoes to England. In 1596, he took part in the sack of Cadiz and helped plan the defeat of the Spanish Armada.

Sir Walter did not fare so well when James I came to the throne; convicted of treason, he was imprisoned in the Tower of London from 1603 to 1616. During this period, he wrote a book entitled *The History of the World*, which was published in 1614. After his release, Sir Walter led an unsuccessful expedition in search of the fabled city of gold known as El Dorado, believed at the time to lie hidden somewhere in the jungles and mountains of South America. In the process of that expedition, he attacked a Spanish garrison, which led to demands for revenge by King Philip of Spain. To appease the Spanish, King James revived the earlier treason charges against Sir Walter who was subsequently beheaded in 1618. The painting *The Boyhood of Raleigh* by Sir John Everett Millais, the Victorian artist and founder of the Pre-Raphaelite brotherhood, was completed while he was staying at Budleigh Salterton, and was set in the vicinity.

Eastwards from Budleigh Salterton lies a stretch of impressive red Triassic sandstone cliffs, huge sea stacks at Ladram Bay and the town of Sidmouth.

Sidmouth

Like Budleigh Salterton, Sidmouth was formerly a port, but has long since silted up and subsequently become a holiday resort graced with fine Georgian and Regency architecture. In 1819, the Duke and Duchess of Kent moved here to escape their creditors, bringing their daughter, the future Queen Victoria, with them. Their former home is now the Royal Glen Hotel. Today, Sidmouth is a relaxing, somewhat old-fashioned resort boasting a repertory theatre, a vintage toy and train shop and museum, a donkey sanctuary and an annual folklore festival. It all reminds me rather nostalgically of the way many seaside towns used to be, as I remember them as a child anyway. One particular item of interest to me in Sidmouth is the Norman Lockyer Observatory, which is at the top of the hill to the east of the town. It was once visible from the town but is now obscured by trees.

The Norman Lockyer Observatory

This observatory is named after Sir Norman Lockyer, who in 1868 discovered the existence of helium in the Sun's atmosphere by means of spectroscopic analysis. Lockyer was also the founder and first editor of the prestigious scientific journal *Nature*. He was knighted in 1897. Today, the observatory is open to the public and displays a range of historic instruments associated with Lockyer's pioneering work on star temperatures.

Norman Lockyer Observatory.

Norman Lockyer Observatory near Sidmouth.

Another more controversial figure associated with the Norman Lockyer Observatory is Donald Barber, who worked there from 1936 until his retirement in 1961. For his final five years at the observatory, Barber held the position of Director Emeritus. Following his death in August 2000, a comprehensive obituary appeared in *The Independent* newspaper, written by George A. Wilkins of the Norman Lockyer Observatory.

The obituary tells us that Donald Barber carried out a photometric investigation of green light from the night sky and was able to confirm a suspected relationship between night sky brightness and geomagnetic activity. He also published papers on seasonal changes in the metabolic activity of both birds and humans and, most controversially, sought to explain the presence of an unknown type of Pseudomonas bacteria found in rainwater used for washing the observatory's photographic plates. He explained it in terms of alien life forms arriving via the solar wind from the upper atmosphere of the planet Venus.

In an article published by the magazine *Perspective* in 1963, entitled 'Invasion by Washing Water', Donald Barber described in detail the regular damage inflicted on photographic plates by these bacteria as they consumed the gelatine on the plates. These bacterial attacks were shown to be cyclic and coincided with a number of factors: an inferior conjunction of the planet Venus (that is when Venus lies directly between the Sun and the Earth), a solar storm, a northerly wind and recent heavy rain. According to Barber, when all these factors coincided

Donald Barber, former
Director Emeritus of the
Norman Lockyer Observatory.
(*Picture kindly supplied by
George A. Wilkins*)

the bacteria appeared. He backed up these claims by publishing his statistical analyses of the phenomenon in the *Perspective* article.

Barber suggested that during a solar storm, when the two planets were in conjunction, material was blown off the top of the Venusian atmosphere towards Earth. He thought that microbes from the upper atmosphere of Venus arrived among that material, following the magnetic lines of force to our polar regions as charged particles upon their arrival on Earth. A northerly wind would bring them south to places such as Sidmouth where, if heavy rain fell, the microbes would be brought down to ground level, in this case into the rainwater used to wash the photographic plates at the Norman Lockyer Observatory.

Having read Barber's article and data, I feel the suggestion is plausible, but the coincidence of all the factors Barber invokes is not always that precise. Dr George Wilkins of the Norman Lockyer Observatory told me that he felt there was probably a local rather than an interplanetary explanation for the mystery bugs.

However, even if Barber's data does not hold up, his idea may yet prove valid. In more recent years, spacecraft have found that the Venusian atmosphere is potentially habitable by some types of extremophile bacteria, and that it does indeed erode in the solar wind, especially during a solar storm. Also, it has now been established that at such times a plume of charged particles from the planet's atmosphere does reach as far as Earth.

Some scientists even claim to have found evidence in the spacecraft data suggesting the presence of microbial life in Venus's atmosphere. In 2008, astrobiologists N. C. Wickramasinghe and J. T. Wickramasinghe published a paper entitled 'On the Possibility of Microbiota Transfer from Venus to Earth'. So who knows, Donald Barber may yet be vindicated and prove to be as important a scientific pioneer as the founder of Sidmouth Observatory, Sir Norman Lockyer.

To Branscombe and Hooken Cliffs

From the heights above Sidmouth, our way takes us on a switchback, roller-coaster of a route along the cliffs towards Branscombe. Here, high, soft, partially vegetated cliffs change gradually from red to white, Triassic mudstone to Jurassic and Cretaceous sandstone and then chalk.

At Weston Mouth, the path descends to the seashore and a beautiful clear stream spilling across the beach to the sea. A number of little huts, some of them apparently of the 'homemade' variety, can be found around here, scattered on the cliffsides or on the beach beyond the high-tide mark. Retreats for some no doubt, they are perhaps the coastal equivalents of the garden shed or allotment for those wanting a 'bolthole' away from their normal domestic surroundings.

Coast east of Sidmouth towards Branscombe.

The shore at Weston Mouth.

Weston Mouth from nearby clifftop.

At Branscombe, we find another stream tumbling seaward, a shingle beach and a welcome café. The village of Branscombe itself lies a little inland and has an attractive thatched smithy dating from the eleventh century. The beach at Branscombe became known across the world after the container ship MSC *Napoli* was wrecked here on 18 January 2007. Crowds from near and far came to look or to take away the goods spilling from numerous beached containers. The contents of these ranged from commercial cargoes, such as wine shipments and BMW motorbikes, to family belongings being shipped from abroad.

Under Hooken and Beer Head

Looking eastwards from Branscombe beach, the white chalk Hooken Cliffs are visible with their tumbled, wooded undercliff below, the result of a massive landslip in 1790 when 10 acres slipped 250 feet. The resultant shelf of debris has since become colonised by scrub and woodland, making this a wild and spectacular scene to walk through once you have passed the beautifully situated static caravans and holiday homes. This scene is made all the more dramatic by the striking chalk pinnacles known as Castle Rock, which now form a second, lower line of cliffs above the sea. This is the first major example on our journey of a coastal undercliff, about which there will be much more to say later.

Hut at Weston Mouth.

The coast near Branscombe.

Caravans beneath Hooken Cliffs.

Hooken Cliffs.

Upon reaching the top of the cliffs known as Beer Head, look back westwards. You will see a large cave (apparently inaccessible) high on a sheer chalk cliff face. This is an adit to the Beer stone quarries, situated a little inland of here. Beer stone is soft when quarried, and hardens on exposure to air. It was used in the construction of Exeter Cathedral among other buildings. The quarries were first dug, it is thought, by the Romans. Today, they are open in summer to visitors and provide an important site for hibernating bats during the winter. Guided tours of the mines are run in the spring and autumn.

The great chalk headland known as Beer Head represents the most southerly and westerly chalk outcrop in Britain. It also has the distinction of being the place of origin of the only piece of chalk to have been sent into space! Strangely, in view of the earlier discussion concerning Donald Barber's findings at the Norman Lockyer Observatory, some of the denizens of Beer Head may have something to tell us about the resilience of microorganisms and the possibility that they could survive transit through interplanetary space. The creatures in question live in clumps of damp moss and on chalk, as well as other surfaces, and are variously known as tardigrades, moss piglets or water bears.

Chalk pinnacles rise out of the green undercliff at Beer Head.

Beer Head with cave just visible and undercliff below.

The Remarkable Water Bear

Water bears, as they are most popularly known, are tiny crustaceans that grow to no more than 0.5 mm in length. They have segmented bodies with eight stumpy legs, which end in little tufts of claws. In drought, they become dormant and roll up into a ball. In this state, water bears can survive almost any conditions. They can be immersed in boiling water or toxic chemicals, subjected to pressures greater than those found at the bottom of the deepest oceans, exposed to intense radiation or left in a complete vacuum. If not physically crushed, they can survive almost anything and then be revived with a single drop of water.

To see how water bears survived conditions beyond the earth's atmosphere in space, some living among moss in the chalk of Beer Head were sent up to the International Space Station in 2008 by the European Space Agency. Here, they were placed outside the space station to endure the vacuum and intense radiation of space. Although some eventually died, most survived in their dormant state and were later revived with a little liquid refreshment.

Beer and Seaton

Moving on from the miraculous life forms to be found at Beer Head, we come to the village of Beer itself. This is a pretty fishing village where you can buy fish directly from the local fishermen whose boats are pulled up onto the beach. It

Axmouth, with Haven Cliffs beyond.

was once famous for lacemaking and the previously mentioned quarrying. In case you were wondering, the name of the village has nothing to do with the alcoholic drink of the same name, but comes from the old Devon word for grove. A footpath past the chalk cliffs leads to the small seaside resort of Seaton.

The feature of most novel interest here is the narrow gauge electric tramway. This 6-mile-long line is operated by the Seaton & District Electric Tramway company, and runs from Seaton to nearby Colyton and back. The open-air tramcars have become popular with bird watchers who use them to get close to the wildlife of the Axe Valley. The tramcars act as mobile hides, as the birds have long since become accustomed to the daily trundling by of the trams.

A short distance to the east is the little port of Axmouth, which in Roman and Medieval times was one of Britain's busiest ports, exporting wool and iron and receiving imports for distribution along the old network of long-distance paths via the prehistoric Harrow Way. Harrow Way linked inland with others such as the Ridgeway and North Downs Way. Axmouth Harbour was eventually blocked by landslides; a local legend explains the occurrence of these in terms of a mermaid's curse on the port following her advances being spurned by a local fisherman who took her fancy.

The towering cliffs here known as Haven Cliffs mark the start of a very interesting and significant stretch of coastline that includes the Axmouth to Lyme Regis undercliff.

Through the Undercliff

The Axmouth to Lyme Regis Undercliff National Nature Reserve, or 'the Undercliff' as it has become known, is mostly visited on foot. Apart from private access about halfway along, the way in and out are at either end of its 6-mile-long footpath. There are no other entrances or exits – once you start, you have to either see it through to the end or turn back!

From Axmouth, the footpath ascends the hill to the east and crosses a golf course before eventually dropping down below clifftop level into a green and jungle-like world of shady woodland, carpeted with hart's tongue fern and offering only the occasional sea view. The footpath is undulating and sometimes muddy or blocked by fallen trees. With no way off once you've started, a long and quite exhausting walk may lie ahead.

The Undercliff was formed by a succession of landslides over many years creating shelves of land between the clifftop and the beach. This land, warm and sheltered, has gradually become colonised by scrub and woodland, but remains an inherently unstable landscape, liable to resume movement at any time, particularly after wet weather. Perhaps the most dramatic landslip within recorded history at this site was one that occurred on Christmas Day in 1839, when 8 million tons of waterlogged chalk slid seawards creating great chasms, isolated chalk plateaux and pinnacles.

Into the undercliff.

Harts tongue fern.

In those days, people lived in cottages in the Undercliff, but in the forty-eight hours or so leading up to the major slip they were all evacuated as significant earth movements had been noticed in the Undercliff itself and just offshore, where reefs had risen as the seabed was pushed up by the forward movement of the land behind.

People flocked to see the result of this landslip and subsequent smaller ones in the following months. Paddle steamer trips were laid on from Weymouth, artworks were produced and a piece of music and dance called *The Landslip Quadrille* was composed to commemorate the event. Some fine and dramatic pictures of the Undercliff after this slip, and the fate of at least one of the cottages following a subsequent slip at Whitlands on 3 February of the following year, can be seen in the Philpot Museum in Lyme Regis, as well as in local booklets and postcards.

More recently, the potential of the Undercliff as a nature reserve has been recognised, and the Undercliff's National Nature Reserve (NNR) was established in 1955. Here, apart from the luxuriant undergrowth, numerous rare invertebrates, including many butterfly species and thirty-three species of wasps and bees, can be found. The woods are largely of ash and hazel, and pyramidal and bee orchids grace open sunny banks. The occasional distant 'roar' you may hear is nothing to be alarmed about; it emanates from roe deer during the rutting season, around July. One natural danger to be wary of, aside from the terrain itself, is the tick.

Uplifted cottage below the February 1840 landslip at Whitlands in the Undercliff. (*Courtesy of Lyme Regis Museum*)

Watch Out – Ticks About!

Ticks are small, blood-sucking parasites that can carry Lyme disease, named, not after Lyme Regis, as you might think, but a town called Lyme in the North American state of Connecticut where it was first identified in 1975. Lyme disease, or encephalitis, starts with flu-like symptoms but can eventually lead to nerve damage. The first danger sign might be a bulls-eye-shaped rash spreading from a tick bite. If you are lucky, you might see the small dark tick before it burrows into the skin, in which case it can be gently removed – though a doctor's visit might still be recommended in case it has infected you. According to Peter Marren and Richard Maybey in their sizeable and entertaining volume *Bugs Britannica*, ticks are not keen on people with red hair. This I can confirm, since my partner, Candida, who has reddish hair, found a tick on her arm after we had walked in the Undercliff. It seemed to have made no effort to burrow into her skin and was easily removed. Apparently, you are more likely to suffer from tick bites in late spring or autumn, and can minimise the chances of a bite by wearing light-coloured clothing and covering up as much skin as possible, for example, tucking trousers into socks.

Hopefully, remaining tick-free, you can enjoy your walk through the Undercliff. Stick to the path and don't be tempted to wander into the woods or look at a pond or reach a sea view, there are many dangerous chasms in the broken ground, obscured by thick scrub.

Left: Pyramidal orchids on a grassy bank, one of eight species to be found on the Undercliff.

Below: Goat Island, an isolated plateau in the Undercliff. (*Photograph by Candida Wright*)

Chapel Rock, a prominent cliff above the Undercliff path. (*Photograph by author and Candida Wright*)

Hidden pond in The Undercliff.

A rare sea view at Pinhay Bay at the eastern end of the Undercliff.

Ruins of a cottage in The Undercliff. (*Photograph by Candida Wright*)

Emerging from the Undercliff at Lyme Regis.

Ammonite paved entrance to the Philpot Museum in Lyme Regis.

You Are Not Alone!

The Undercliff has naturally had some impact on locally set literature. In Jane Austen's *Persuasion,* we read of its 'green chasms between romantic rocks'. More recently, local author John Fowles, in *The French Lieutenant's Woman,* used the Undercliff as the place where the wayward Sarah Woodruff was firmly and quite clearly told by her strict employer Mrs Poultenay *not* to go walking, since it would be considered 'unseemly' on account of 'what goes on there'; it was popular with courting couples. That was in 1867. Today, if you and your companion(s) find yourselves alone in the Undercliff, just sit down for a few minutes and perhaps have a drink from your flask of tea – just see how many people stride past you! When you keep moving, the illusion of being relatively isolated in a wilderness is maintained. Sit down for five or ten minutes and the truth is revealed!

Eventually, the Undercliff footpath emerges into Lyme Regis, having just crossed the border from Devon into Dorset. If you turn right when reaching the town, the beach you find yourself on after a little while is known as Monmouth Beach, because it was here that the Duke of Monmouth landed in 1685 to mount his ultimately unsuccessful attempt to remove James II from the throne. This beach is also known as one of the best places anywhere to find large fossil ammonites, which are often visible for all to see in fallen rocks, but more on fossils and Lyme Regis in the next chapter.

2

Lyme Regis to Chesil Beach

'The Pearl of Dorset' and 'the Pearl of the Jurassic Coast' are just two enthusiastic and entirely deserved descriptions of Lyme Regis. This wonderful little town spills down narrow streets to the sandy beach and picturesque harbour with its curved breakwater known as the Cobb. This has been immortalised through literature and film as the place where Sarah Woodruff watches hopefully for her lover to return from across the sea in *The French Lieutenant's Woman*. The streets are full of delightful cafés, bookshops, fossil shops, an old mill, seaside pubs, a cannon overlooking the sea, the fascinating Philpot Museum with its entranceway paved with large fossil ammonites, the Dinosaurland fossil museum, airy promenades and glimpsed views of the scenic delights ahead.

Bustling Lyme Regis seafront with pleasant gardens behind.

View from a seafront café with cannon pointing the way to Golden Cap.

View from busy Lyme Regis Harbour towards Golden Cap.

The town gained its 'Regis' title when it was granted a royal charter by Edward I in 1284, having been charged with supplying ships for the King's navy in a similar arrangement to that of the Cinque Ports of the Kent and Sussex coasts. Today, the harbour is only used by local fishing boats and yachts. Fresh fish are landed daily, tides and weather permitting, and boat trips for angling purposes or to view the Undercliff from the sea can be taken from here. Lyme Regis also boasts what is claimed to be the oldest post box in Britain, distinguished by its two posting slots, one horizontal and one vertical – the latter used for those posting from horseback.

Mary Anning, Fossils and Mudslides

It is fossils that Lyme Regis is most well-0known for today. This is courtesy of a local cabinetmaker's daughter, Mary Anning, who in 1811 at the age of twelve made the first discovery of a complete fossil skeleton of a 21 foot-long Icthyosaurus. Subsequently, she uncovered many other notable fossils, including a Plesiosaurus and a Pterosaur among the rocks below Church Cliff and Black Ven, which can be seen to the east of the town. Mary Anning is said to have been the inspiration behind the well-known tongue twister, 'she sells seashells by the seashore'. She gained a formidable reputation for her fossil collecting and academic research, but, being a 'mere woman', was not allowed to join any of the professional scientific organisations or societies at that time. As the years went on, she fell into financial difficulties but managed to secure a pension, which gave her some financial security in her declining years. Sadly, Mary Anning died at the relatively young age of forty-seven as a result of breast cancer. Appropriately, her former home is now the excellent Philpot Museum in Lyme Regis.

Coincidentally, while writing these words, I have just heard that today (6 January 2014) another complete Icthyosaurus skeleton has been uncovered at Black Ven following yesterday's severe winter storms. Black Ven, along with Monmouth Beach, are the prime places in Britain to search for the fossilised remains of the abundant marine life of the Jurassic and Cretaceous periods. You can, of course, find fossils all over Lyme Regis, on display or for sale, and see some of the more spectacular and significant Mesozoic reptile remains at the Dinosaurland fossil museum. There's nothing like finding your own though, as many people do throughout the year, particularly at weekends or after high seas have uncovered more at the foot of the cliffs. If you do go fossil hunting, all the usual caveats apply. Only pick up from the beach. Don't dig into the cliffs. Know the tides so as not to get cut off beneath the cliffs with no escape. Watch for falling rocks and don't try to cross mudslides, as soft mud often traps the unwary. You could even drown in a deep mudflow!

This latter problem is a particular issue below the grey clay and shale cliffs of Black Ven, where in the winter of 1957/58, the largest mudslide in Europe occurred after heavy rain. Smaller mudflows happen every winter, so the danger is always there.

The cliffs of Black Ven can be seen beyond the town.

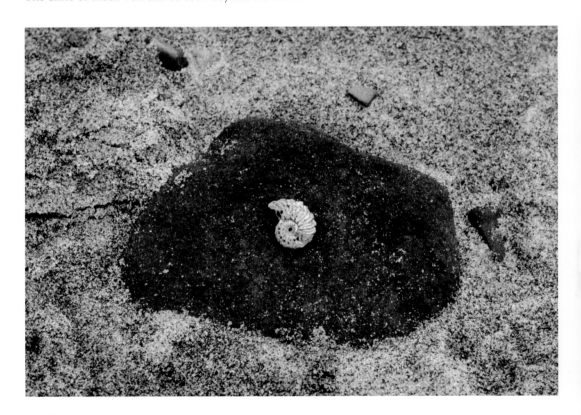

Fossilised Cretaceous ammonite.

Leaving Lyme Regis behind. (*Photograph by Candida Wright*)

Looking west across Undercliff landscapes towards Lyme Regis, from above Black Ven.

Above: Blue Lias landslips and the 161-acre Black Ven (the Spittles nature reserve), west of Charmouth. (*Photograph by Candida Wright*)

Left: Church of St Candida and Holy Cross at Whitchurch Canonicorum, near Morcombelake. (*Photograph by Candida Wright*)

One particularly interesting mudslide occurred in 1908 when the action of the collapsing cliff at Black Ven set light to a band of bituminous shale. Smoke billowed up from a conical formation surmounted by a hole, leading to reports of a 'Lyme Volcano'. When investigated at first hand, the inside of the hole was found to have a 'baked' appearance. We will encounter more reports of similar phenomena as our journey along the Jurassic Coast continues.

An Ecclesiastical Diversion

Avoiding the mudflows and landslips of the shore, the footpath follows the clifftops eastwards, but in recent years has become seriously detoured inland, due to severe coastal erosion, and reaches the sea again at Charmouth. If following the road eastwards, you may like to explore the village of Morcombelake. This was where I found a room for the night way back in 1974 on my first ambitious attempt to walk from Sidmouth to Swanage in one go, though I was then completely unaware of the secrets of the nearby countryside. The pub that offered me a room has long gone, but the secrets are still there to be found. I refer to the church of St Candida and Holy Cross and St Wite's Well that lie in the vicinity.

Heading east, the road to St Wite's Well is on the right as you pass through Morcombelake. This takes you to Chardown Hill and the National Trust maintained well, which has long been associated with curing sore eyes. It is believed locally that St Wite was an Anglo Saxon, a Christian and a hermit, who lived on Chardown Hill in the vicinity of the well. She was martyred along with other local Christians when a band of some 15,000 Vikings (could it have really been that many?) landed on the coastline at Charmouth. Alfred the Great is credited with establishing a church of St Wite at a nearby village, now called Whitchurch Canonicorum, some fifty years after the Viking raid.

The Bay & Burning Cliff, Lyme Regis.

Contemporary postcard view of the smoking 'Lyme Volcano', photographer unknown.

This church can be found by taking either of the left turns out of Morcombelake when heading eastwards. Today, it is known as St Candida and Holy Cross. At some point in the Middle Ages, the name 'Wite' became synonymous with 'white', which, when translated into Latin, became 'candida'. Between 1200 and 1505, 'St Wite' seems to have become 'St Candida', at least as far as the church is concerned.

The relics of this saint are kept within the church, which isone of only two parish churches in England to still have the remains of their founding saint. The other is St Eanswythe's church in Folkestone, Kent. Of course, Westminster Abbey holds the body of Edward the Confessor.

According to the leaflet *The Cathedral of the Vale*, available in St Candida and Holy Cross church, this holy site became a pilgrimage destination second only to Canterbury during the Middle Ages. Pilgrims still come and healing is still sought, with many tourists now visiting the shrine within the church. A statue of St Candida can be seen in a niche high up on the outside of the church. This was given to the church in 1980 by Dorothy Sims-Williams.

On a personal note, my first attempt to find St Wite's Well failed, and my partner Candida (yes, that is why we were here following this up) and I ended up at the church. We decided to look for the well another day as I had been suffering

Shrine of St Candida at St Candida and Holy Cross, Whitchurch Canonicorum, near Morcombelake. (*Photograph by Candida Wright*)

Statue of St Candida in
a niche on the outside of
the church at Whitchurch
Canonicorum, near
Morcombelake. (*Photograph
by Candida Wright*)

from sore eyes. On returning to Lyme Regis, I discovered that I had lost a
lens from my spectacles somewhere en route and had to buy a new pair. After
this, the problem with the sore eyes disappeared and I concluded that the cause
of the ailment was that my previous glasses had been of the wrong magnification.
So, my sore eyes were cured! What are we to make of such coincidences? It seems
I did not need to bathe my eyes in the waters of St Wite's Well to effect the
desired cure; the visit to the church and the shrine alone had done the trick. I am
by no means religious but it does bring to mind the words of William Cowper's
hymn: 'God works in a mysterious way, his wonders to perform'. Of course,
Buddhists might say this sort of thing points to the dreamlike nature of reality,
while Jungian psychologists might point to 'synchronicity' without any idea how
this might actually work. Suffice to say, it's a mystery.

Stanton St Gabriel

From Morcombelake, a path leads back to the coast through fields and woods,
past a pond, over a stream and through the hamlet of Stanton St Gabriel. In the
seventeenth century, twenty-three families lived here, but now there are just a
couple of buildings used as holiday lets and a small, picturesque, ruined and
roofless thirteenth-century church.

Remaining houses in Stanton St Gabriel. (*Photograph by Candida Wright*)

Ruins of the thirteenth-century church of Stanton St Gabriel. (*Photograph by Candida Wright*)

View west from the ruins of the church of Stanton St Gabriel.

No service has been held at St Gabriel's since the late eighteenth century, and during the Napoleonic Wars it was used as a receiving house for smuggled brandy and other contraband. At the time of my last visit, the arched doorway into the ruined church played host to a large and menacing hornet that inhabited a hole in one of the archway walls. This made for a rather alarming entry and exit to this intriguing ruin.

To Golden Cap and Beyond

Back on the coast, the scene is one of wonderful, undulating downland rich in wildlife and with cattle grazing, mostly owned and managed by the National Trust. St Gabriel's stream tumbles through a ravine to the sea, while the coastline itself shows signs of having slid and tumbled into more undercliff landscapes, both at Cain's Folly towards Charmouth and eastwards to the mountain-like and mysteriously flat-topped Golden Cap.

Golden Cap is so called because of its sandstone summit, which looks golden in the evening sun. It is the highest point on the Jurassic Coast and indeed anywhere on the south coast of England. Just how high it is seems disputed. The guidebooks variously give Golden Cap's highest point as 618 feet, 619 feet, 626 feet and 627 feet. When metric measure is used, however, everyone seems to agree that the cliff's height is 191 metres, which is what the Ordnance Survey map shows, so I think we'll leave it that.

Towards Lyme Regis.

Landslips near Golden Cap. *(Photograph by Candida Wright)*

View west from the top of Golden Cap.

The way up Golden Cap from the east.

The National Trust own nearly 2,000 acres and 5 miles of the coastline around here, which is known as the Golden Cap Estate. This was acquired in stages between 1961 and 1994. Just why Golden Cap is so flat on top is hard to say. Obviously, it has been artificially flattened at some point, and being the highest point around it makes an ideal lookout or signalling location. I understand that the flattening may have taken place early in the nineteenth century as part of the preparations for the expected invasion by Napoleon's forces, which of course failed to materialise.

At the summit of Golden Cap there is a memorial to Lord Antrim, chairman of the National Trust from 1965 until his death in 1977. There are, as you might expect, the most stunning views along the coast in both directions.

To the east of Golden Cap, the cliffs descend in height gradually, from the crumbling heights of Doghouse Hill and Thorncombe Beacon to the sheer sandstone drops at West Bay and the low cliffs of Burton Beach. Behind the coastline of Golden Cap and Doghouse hill lies a countryside laced with hidden, sunken footpaths known as 'holloways', created by many centuries of footfall. These occur throughout the country, wherever the ground is soft enough to wear away like this and has not been subsequently developed or farmed.

Robert Macfarlane, with his friend and fellow author Roger Deakin, explored the 'holloways' of the south Dorset countryside. Macfarlane repeated the experience six years later with friends Stanley Donwood and Dan Richards, inspired by Geoffrey Household's book *Rogue Male,* which was published in 1939 and told the story of a pursued man who had 'gone to ground' in this area among these lost, hidden features of the landscape. Their journey has been beautifully recreated in words and wonderful line drawings of the scenes within such overgrown 'holloways' by artist Stanley Donwood in a slim volume called simply *Holloway,* which was published in 2012 as a limited edition and more widely in 2013. Interestingly, the authors say in their introduction that a map of the holloways finding is 'not contained within it', which I thought was a rather good decision.

Recently, this particular stretch of coast has become popular with film-makers. Scenes from a new film version of Thomas Hardy's *Far from the Madding Crowd,* due for release in 2014, were filmed on downland near West Bay, while the television drama series *Broadchurch,* starring Olivia Colman and David Tennant was set in West Bay itself. The writer and creator of *Broadchurch,* Chris Chibnall, has been quoted in a local tourist guide as saying that the drama 'was written as a love letter to the Jurassic Coast'.

West Bay is actually quite a modern-looking, though modest, seaside resort, formerly known as Bridport Harbour. Bridport itself, the 'Port Bredy' mentioned in some of Thomas Hardy's novels, is a little inland now, but from the thirteenth century it was well-known as a centre for rope making. The town's harbour, now called West Bay, was built in 1740 and catered for up to 500 ships a year in the nineteenth century. Various ships, including naval vessels, were built here until 1879. Today, it is used mostly by fishing boats and yachts.

Looking towards Doghouse Hill.

East of West Bay, golden sandstone cliffs line the coast towards Burton Beach. When the setting sun illuminates these cliffs, it is easy to see why West Bay has sometimes been called the 'golden gateway to the Jurassic Coast'. Solid as these cliffs appear, they are no safer to walk beneath than any others along the Jurassic Coast, as was tragically proved in 2013. After Burton Beach, the cliffs peter away and shingle beach becomes the dominant coastal landform, we have now reached Chesil Beach.

Along Chesil Beach

This shingle bank stretches for a total of 18 miles towards the strange tabletop landform of the Isle of Portland, which is visible ahead. No one is sure exactly how this bank is formed and maintained, but it is certainly a product of longshore drift and is composed of pebbles from as far away as Budleigh Salterton. These arrange themselves by size with the smallest at the western end and the largest to the east towards the Isle of Portland. Indeed, it is said that these pebbles are graded so precisely that a knowledgeable local mariner washed up on a foggy night anywhere along the coast from Lyme Regis to Portland would know just where they were by the size of the stones alone.

For 9 miles, Chesil Beach runs parallel with, but separated from, the rest of the shoreline by a shallow strip of brackish water known as the Fleet. This is rich in eel grass, which is popular with many seabirds and water fowl as well as the large numbers of mute swans that are to be found near the picturesque village of Abbotsbury. The swans were possibly introduced in the eleventh century by Benedictine monks at the monastery at Abbotsbury to provide a source of fresh meat, though a definite record of them only dates from 1393. Ownership of the monastery, the Fleet and the surrounding land passed to the Fox-Strangeways family in 1543 and has remained with them ever since.

The Swannery, as it is known, is now a popular tourist attraction. Today, these swans are said to be the largest managed herd in the world; about 600 feed here daily. Apparently, since the 1740s, feathers from the swans have been used by Lloyds of London for writing in the *Doom Book*, their record of ships lost at sea.

Above Abbotsbury on a prominent hilltop stands the imposing St Catherine's chapel, one of the iconic views of the Jurassic Coast, seen to best effect from the top deck of the X53 bus when approaching Abbotsbury from the west.

Further along the shoreline opposite Chesil Beach is the small village of East Fleet. Most of the earlier village was swept away in a tidal surge in 1824. The village of East Fleet was the inspiration behind the fictional village of Moonfleet,

Abbotsbury swans with Chesil Beach seen on the horizon. (*Photograph by Candida Wright*)

Abbotsbury swans. (*Photograph by Candida Wright*)

Shoreline of the Fleet. (*Photograph by Candida Wright*)

Above: muddy shoreline and boat by the
Fleet, with Chesil Beach in the distance.
(*Photograph by Candida Wright*)

Left: Chesil Beach seen from the
Isle of Portland.

which featured in J. Meade-Falkner's absorbing tale of eighteenth-century smuggling on the Dorset coast, entitled *Moonfleet*. A large house at East Fleet, built in 1603, featured in the story and is today the Moonfleet Hotel.

Beyond Abbotsbury, the long-distance footpath can take you either along the shore-side bank of the Fleet or up on to the green, tumuli-peppered downland behind. An Iron Age hill fort behind Abbotsbury known as Abbotsbury Castle offers particularly fine views both east and west. The main natural features visible ahead are Chesil Beach and the Isle of Portland beyond.

Walking Chesil Beach is not recommended, as 9 miles of slogging through pebbles that become larger the further east you go can be very hard work. Swimming off Chesil Beach is also not recommended due to unpredictable currents. Roger Deakin, in his book *Waterlog*, tells how well-known author Iris Murdoch had a narrow escape while swimming here.

Author Ian McEwan makes good use of the dramatic scenery of Chesil Beach in the emotional finale to his novel *On Chesil Beach,* which tells an insightful and thought-provoking story of fear, expectation, disappointment and lost opportunity in a doomed 1950s marriage. It is a tale that ends with the consequences of things left unsaid and their impact on the lives of the main characters. It is a tale that leaves us wondering, perhaps in a personal sense, 'if only'.

I am left wondering, too, what it is about the Dorset countryside and coastline that inspires writers to set rather moody, sometimes gloomy tales of love and loss here?

Ahead now, beyond Chesil Beach, the prominent, grey tableland of the Isle of Portland beckons. This is the subject of our next chapter.

Below: Chesil Beach, image *c.* 1890. (*Courtesy of the Library of Congress*)

3

Around Portland

The Isle of Portland is not an island as such but more of a peninsula connected to the 'mainland' by a narrow neck of land along which the main road runs. It was used as the setting for two of Thomas Hardy's novels, *The Well Beloved* and *The Trumpet Major,* and described by him as the 'Gibraltar of Wessex', on account of its appearance as a distinctive promontory. He also calls it the 'Isle of Slingers', a reference to the long-established stone working industry of the 'island'. Stone is what Portland is all about.

This treeless, windswept plateau has constant reminders of stone working everywhere. Wherever you look there are quarries, in use or abandoned, caves where stone has been excavated from cliff faces, piles of rubble, sharp edges, fissures and cracks.

Many writers have not been kind to the Isle of Portland. In 1958, film-maker John Read described it as having an 'abandoned air', saying that 'a litter of isolated objects lie about, scattered over the shore and landscape. Rocks and lobster pots. Old boats on a deserted coast ... stones piled high like toy bricks ... a surface pitted with disused quarries, huts and heaps of rubble.' Likewise, Sir Frederick Treves, who is probably best known as the London surgeon who helped the famous Elephant Man 100 years ago, informed us that the flora of Portland consists of 'coarse grass, a few teazles and an occasional starving bramble bush' in his book *The Highways and Byways of Dorset.*

Some artists, however, are drawn to such landscapes. Having visited and painted the similarly bleak landscape of Dungeness in Kent, John Piper came here in 1929 and returned in 1934 having found plenty of material to work with. Some of his paintings show the tumbled stone blocks and deserted landscapes mentioned earlier, as well as the Portland lighthouse and much sculptural detail of the St George Reforne church at Easton, though the latter paintings were completed some twenty years later. One of John Piper's few drawings to include a human presence was of Chesil Beach after an exhausting walk along the shingle bank with his wife Myfanwy. The drawing, entitled simply 'Chesil Beach', depicts his wife resting on the shingle, among the flotsam and jetsam, at the end of their walk.

Without doubt, the Isle of Portland presents a harsh prospect for most people. It is gradually returning to nature but was formerly an almost completely industrial landscape. The cause of all this stone working is the nature of Portland limestone, which is a fine, white limestone that is easy to cut and carve but hardens as it oxidises on exposure to the air. There are a number of quarries on Portland but perhaps the most interesting is Tout Quarry, situated atop the precipitous cliffs of West Weares.

Tout Quarry

Quarrying started at Tout Quarry in Roman times, and stone from here was used to build the Tower of London in the fourteenth century. After the Great Fire of London in 1666, the architect Sir Christopher Wren, who had also been an MP for Weymouth, decided upon Portland stone as a suitable material for the great rebuilding process that began in the wake of the fire. Many important buildings, both then and subsequently, have had Portland stone used in their construction, namely, St Paul's Cathedral, Buckingham Palace, the Bank of England, the British Museum and, in New York, the headquarters of the United Nations.

Commercial quarrying at Tout Quarry ceased in 1982, but sculptural artists soon began to make use of the abandoned stone to practice their craft. The Portland Sculpture and Quarry Trust was created in 1983 to preserve knowledge and promote understanding of Portland stone and the landscape from which it comes. An active workshop operates within the quarry that caters for artists at all levels and colleges and schools and runs courses and workshops for the public all year round. Sculptors can carve their creations in the quarry and leave them there in their place of origin for all to see. Such famous names as Sir Antony Gormley, Henry Moore and Barbara Hepworth have all used stone from Tout Quarry. One particularly 'fluid' and well-photographed sculpture on outdoor display called *Falling* is a creation of Sir Antony Gormley. The outdoor workshop usually runs from May to September. Tout Quarry is also the site of a nature reserve run by the Dorset Wildlife Trust. Many types of lichen and the rare chalk hill blue and Adonis blue butterflies can be found here.

Moving up the windswept, sea-battered west coast of Portland brings us to the Bill of Portland, or Portland Bill as it is usually called. This is a notoriously dangerous stretch for small boats on account of a tidal race called, not unsurprisingly, the 'Portland Race'. I once waited for hours at Lymington in Hampshire for my sister and her husband to sail round from Lyme Regis, despite a favourable wind. I remember the reason for their late arrival being 'whirlpools' encountered off Portland Bill. This is apparently caused by a clash of tides and currents around the promontory.

A large rock just offshore, known as Pulpit Rock, may look familiar, as it is one of the most photographed of Portland scenes and often appears in guides and leaflets on the Jurassic Coast. This is not, however, an entirely natural formation. It was once a stone archway created by erosion from the sea but this was demolished by quarrymen, leaving just the huge stack we see today. Nature, of course, also does its share of demolition. A similar offshore rock stack known as

Stone carving at Tout Quarry. (*Photograph by Ray Hollands*)

Stone carving at Tout Quarry. (*Photograph by Ray Hollands*)

Stone carving at Tout Quarry. (*Photograph by Ray Hollands*)

Pulpit Rock with fisherman atop. (*Photograph by Maureen Barkworth*)

Pom Pom Rock was smashed and totally destroyed by waves hitting the Portland coastline during the severe storms of January 2014.

Portland Bill itself has a tall lighthouse, built in 1906, which is now fully automatic. The former lighthouse keeper's housing next to the lighthouse was consequently made redundant, though it is now put to good use as the Portland Bill Lighthouse Visitor Centre. Here you can learn all you ever wanted to know about the Isle of Portland and watch a live video link covering the comings and goings of birdlife on the nearby cliffs.

Church Ope Cove

Following Portland's east coast, back towards Weymouth, a further stretch of 'quarry cliffs', which include a cliff with an impressive cave and blow hole, give way eventually to the 'island's' only beach, at Church Ope Cove. The site of the beach is marked by a castle ruin, from which a path passes through an archway and down a steep flight of steps to reach the sheltered beach. This is considered by many to be the most attractive part of Portland's coastline. A hidden path known by locals leads through a cleft from the steps to what has been described as a 'Roman' reservoir, though others say it is Victorian. This sounds like something intriguing to explore, although I personally have not done so yet. I can only surmise that this has something to do with the nearby castle ruin and so is probably medieval or later.

The ruined castle itself is gradually succumbing to erosion and falling into the sea. This is known as Rufus Castle, or 'Bow and Arrow Castle' to some. 'Rufus' refers to William II, for whom the castle was originally built. His red hair earned him the nickname 'Rufus'. Another interesting snippet relating to Church Ope Cove is the old tale of a mermaid being rescued from drowning in the waters off the cove many years ago. She was apparently carried up to the nearby St Andrew's church but soon died. It is hard to know what to make of such tales, especially when they are alleged to be true. I have heard a plausible suggestion that before sea bathing and swimming for pleasure became popular, it was almost unheard of, especially as it was likely that hardly anyone knew how to swim anyway. So any occasional young woman seen swimming in a remote spot like this might be considered quite unusual and noteworthy – possibly a mermaid! The story presumably relates to a period long ago, as St Andrew's church, the oldest building on Portland, has not been in use since the eighteenth century and is now an overgrown ruin. Leaving the intriguing ruins and attractive Church Ope Cove behind us, we head north past a scrubby undercliff and to a view of Portland Harbour.

From Durdle Door to Bats Head. (*Photograph by Maureen Barkworth*)

Portland Harbour

One of the largest man-made harbours in the world dominates the scene as we move to the north of the Isle of Portland. Started in 1849, it took twenty-three years to build and prison labour from nearby Verne Prison was used in its later stages. In June 1944, Portland Harbour saw the departure of the United States Infantry Division for the beaches of Normandy during the D-Day invasion of Europe. The harbour remained an important naval base until the 1990s when it was developed by a private company into a successful deep water port.

The nearby Portland Castle, in the village of Fortuneswell, is one of Henry VIII's extensive string of fortifications built in 1538–40, which includes Sandsfoot Castle across the water in Weymouth. Bringing the story right up to date, Portland Harbour and Weymouth Bay hosted the sailing events for the 2012 Olympic and Paralympic Games. A modern training facility run by the National Sailing Academy has, among other improvements, provided a lasting legacy for Portland and Weymouth in the wake of the Games.

Across the bay from Portland, Weymouth and the line of cliffs towards Lulworth Cove entice. They are the subject of our next chapter.

4

Weymouth to Lulworth Cove

Until 1789, Weymouth was little more than a fishing village. Then came George III who became the first monarch to use a bathing machine on Weymouth beach. The place subsequently became a fashionable resort, as is evident from the fine Georgian architecture to be found in the town. From 1789 to 1805, George III had a summer home in Weymouth at Gloucester Lodge, which has since become the Gloucester Hotel. The King's former presence in Weymouth is marked by a fine statue of him overlooking the promenade. This was erected to mark his Golden Jubilee. A couple of miles along the coast, cut into the hills behind Osmington, a huge chalk figure of George III on horseback can be seen.

Weymouth has continued to be a successful holiday resort. A long promenade curves around a beautiful sandy beach upon which can still be seen the traditional seaside, the children's entertainment show Punch and Judy, as well as some amazing sand sculptures. There are all the usual facilities of a medium-sized coastal resort, including an impressive sealife centre with a futuristic observation tower that allows stunning views along the Jurassic Coast. The centre boasts over 1,000 marine creatures, including sharks, seals, spider crabs, octopuses, piranhas, sea turtles, otters, crocodiles, axolotl, sea dragons and more. Another interesting wildlife attraction lies slightly inland at Radipole Lake, where the RSPB visitor centre has underwater webcams showing what is going on beneath the lake's surface. Weymouth has benefited greatly from the 2012 Olympic and Paralympic Games, and exudes an air of prosperity lacking in many British seaside resorts today.

Some of the most beautiful coastline in Britain lies not far east of Weymouth. You can drive a couple of miles along the coast to Bowleaze Cove, where the road turns inland. Here there is parking, an entertainment centre, funfair and access to the beach. The coast rises up into low cliffs and the coastal path presses on to Osmington Mills. At nearby Jordan Hill there are the foundations of a Romano-British temple of the late Roman period, around AD 390–420. An illustrated panel gives more information and has a pictorial representation of what the temple may have looked like when it was still standing.

Osmington Mills

This small hamlet is mostly concerned with providing a pleasant family holiday in beautiful surroundings by the sea. Low-key camping and self-catering facilities do not spoil the coastline and have easy access to the coastal path, rock pools on the beach and a small entertainment complex. The Smuggler's Inn public house also provides bed and breakfast and a nice sunken garden where supping a drink on a warm evening by a calm, moonlit sea has an exotic feel. A flight of steps leads down to a jetty by the sea and the cliffs are low and luxuriantly vegetated. There is something quite paradisical about this spot, the stuff of summer holiday memories.

Spontaneous Combustion at Burning Cliff

Striding on brings us to another slightly larger village at Ringstead Bay. Beyond this, the land can be seen rising steeply into a line of 400-foot-high cliffs and the chalk headland of White Nothe with an interesting tumbled undercliff beneath. Ascending the cliffs via the coastal path, you pass across an area at the start of the undercliff known as 'Burning Cliff', so called because of an interesting phenomenon that started here in 1826 and continued for some years.

A full description of this phenomenon is given in the book *Geology of Weymouth and the Isle of Portland* by Robert Damon, published in 1880. In this, we read of an occurrence similar to the one that later caused the 'Lyme Volcano' mentioned in Chapter 2. Damon explains,

> In Holworth Cliff, Ringstead Bay, spontaneous combustion proceeded in the autumn of 1826 and continued for some years. Sulphurreted hydrogen was liberated, the odour of which could be detected for several miles under certain circumstances.

Damon noted that a similar ignition had taken place at Charmouth in 1751 and again in 1755. He goes on to explain the process:

> This phenomenon first made its appearance immediately after a spring tide, which was being attended with strong southerly gales, the water producing decomposition of the iron pyrites in the shale was supposed to have been the cause of the ignition.'

Damon explains that earlier the cliff had slipped, exposing a fresh portion of it to the wind and waves. The following account describes what apparently occurred:

> The cliff continued in this burning state for several years during which period it formed an object of considerable interest. On acquiring fresh energy, it threw out volumes of dense and suffocating smoke, which from its specific gravity, seldom rose high in the air. This was followed by bluish flames rising at times so far above the

Unspoiled coastline near Hannah's Ledge.

Ringstead Bay with cliffs beyond.

Above: Holworth Cliff with Holworth House above the area known as 'Burning Cliff', where spontaneous combustion took place in 1826.

Right: A path down to the beach at Osmington Mills near Hannah's Ledge.

A tired walker leaving Ringstead Bay and Holworth House behind on the way to White Nothe.

cliff as to be visible from Weymouth. Through the cracks spread over the surface by the ascending heat, the burning stratum beneath was seen. The fissures and other openings were covered with deposits of sulphur.

Damon goes on to describe the discussion about the commercial prospects of using the ignitable shale oil in the cliffs as a fuel source, which was not considered viable. That was then. Now, shale oil is of great interest as it can be extracted by the controversial process of 'fracking'. Much of the Dorset coastline and countryside is of particular interest to those seeking to carry out exploration with a view to mining in the future. We can be assured, however, that though this may occur inland in the future, the Jurassic Coast now has considerable recognition and legal protection that should ensure its preservation from the ravages of future mining.

White Nothe Undercliff

Moving higher up the cliff path, we leave Burning Cliff behind and mount the cliffs to White Nothe and its green hummocky undercliff. The stretch of coastline from Ringstead Bay to White Nothe is owned by the National Trust, having been acquired in stages between 1949 and 1984. The White Nothe undercliff covers about 115 acres and is a landscape of miniature hillocks and valleys dotted with blocks and pinnacles of chalk, resembling ancient megalithic structures or the

The headland of White Nothe with undercliff below and old coastguard cottages just visible up on the clifftop.

ruins of old cottages, beneath partially vegetated high chalk cliffs. A rock and shingle beach lies at its foot. This undercliff was of course created by landslips, as is the case with the coastline either side of Lyme Regis or at Under Hooken near Branscombe. In this case, it is a much more open, grassy landscape that is home to a variety of insects, many wild flowers and adders that enjoy basking in spring sunshine.

There are two paths down to the White Nothe undercliff from the coast path above. Take the first one, a pleasant walk down towards the shore, for the later path at the headland of White Nothe itself, which descends from near the coastguard cottages, is precipitous in the extreme. This vertiginous footpath was used as the setting, along with the undercliff itself, for dramatic scenes in J. Meade-Falkner's earlier mentioned smuggling tale, *Moonfleet*. He refers to White Nothe as 'Hoar Head'.

To Durdle Door and Lulworth Cove

It is not without reason that these few miles are constantly featured on calendars, in books, leaflets and indeed printed or photographic matter of any kind relating to Britain's coastline. This must be not only the most photographed part of the Jurassic Coast, but of the British coastline as a whole. The walk from White Nothe to Lulworth Cove will 'wow' the most jaded walker. This could justifiably be called the finest cliff walk in Britain. There is a certain kind of majesty, brightness and warmth that only chalk cliffs exude. These (mostly) chalk cliffs also display a remarkable variety, each one different from the one before or after, a truly stunning piece of coastal geomorphology.

From White Nothe, the coast path descends almost to sea level, though not as precipitously as with the path down to the undercliff, then up again very steeply. In fact, this is quite a roller coaster of a route, exhausting and time-consuming for just a few miles but scenically rewarding in the extreme.

At Bat's Head you can see a 'baby' version of what is ahead. It has been called a 'mousehole', a young, still small, natural archway that will gradually enlarge to resemble Durdle Door, which is still ahead, a little to the east. Sometimes you can see brilliant metallic green rose chafers, an almost luminescent flying beetle that usually

Old drawing from the book *Weymouth and the Isle of Portland, c.* 1880, looking west across 'Durdle or Barn Door Cove'.

Durdle Door. (*Photograph by Candida Wright*)

flies singly and the similar but brown-winged cockchafer that sometimes flies in great swarms. Many years ago, my walking companion and I were attacked by such a swarm that literally flew straight at us, falling to the ground on impact. What prompted such behaviour I cannot say. Perhaps we were just in their way and their collision with us accidental. We were unharmed, but I'm sure the beetles were less fortunate.

Approaching Durdle Door there is a point where you may be forgiven for thinking that the path has veered off in another direction, has collapsed or just inexplicably ended. I refer to the precipitous 'Scratchy Bottom'. Here, I found myself having to descend quite alarmingly down an almost a vertical incline. Thankfully, it was partially covered in grass and earlier walkers had left an only just visible trail of flattened 'steps', or so it was when I first walked it. Now, the rather dangerous sections of the coast path have been improved with proper steps and fences to ease the way for the thousands of walkers who now follow this route. As for the amusing name, I can only assume this was descriptive of what happened to you if you inadvertently descended the precipice too quickly.

Finally, the huge natural limestone arch of Durdle Door comes into view, that iconic image of the Jurassic Coast known the world over. I first saw it all those years ago in the opening scene of the 1967 film version of *Far from the Madding Crowd*. It made another appearance

Another view of Durdle Door. (*Photograph by Maureen Barkworth*)

further on in the film as the place where Sergeant Troy, played by Terence Stamp, faked his drowning. At this point, it is worth looking back along the coast towards White Nothe for a truly stunning view that is instantly recognisable from many book and calendar covers.

To reach the beach at Durdle Door requires descent of a steep flight of steps, but once there you will be rewarded with crystal-clear waters full of fish and marine life, excellent for snorkelling or underwater photography. Less energetically of course, you may prefer to sit and enjoy the incredible scenery.

Beyond Durdle Door, more exhausting ups and downs, probably in the company of a great many other walkers, bring you past the contorted strata of the dramatic Stair Hole where a new arch and cove are now forming, to the attractive Man O' War Cove, a precursor to Lulworth Cove, that other big geological celebrity of the Jurassic Coast.

Lulworth Cove was carved out when the sea broke through a line of harder cliff material and eroded the softer chalk behind. The process is no doubt still in progress. One day in the future, Stair Hole may resemble today's Lulworth Cove. The natural coastal amphitheatre of Lulworth Cove is extremely popular with visitors. There are boat trips out from the cove and an interesting heritage centre that tells the geological history of the cove and the nearby coastline. The naturalist may be interested in the rare brown and black butterfly that takes its name from Lulworth, the Lulworth Skipper, which you may see fluttering above the grassy clifftops anywhere from Sidmouth to Studland.

On the opposite (eastern) side of Lulworth Cove, the land is mostly out of bounds except at weekends, as this forms part of the boundary with an Army training area covering some 7,000 acres in all and includes the 'ghost' village of Tyneham, but we will cover this in the next chapter.

Below: Towards Lulworth Cove. (*Photograph by Maureen Barkworth*)

5

West Lulworth to St Aldhelm's Head

From the coast just east of Lulworth Cove to Kimmeridge 6 miles away, the footpath is closed except at weekends and some holiday periods, so choose when you go. The Heritage Centre at Lulworth Cove and other information and visitor centres along the Jurassic Coast should be able to advise you. The Army firing range is visible as you walk along the path, though of course military exercises do not take place when the path is open to walkers.

One of the first points of interest is the Fossil Forest, reached by a flight of steps down the cliff side. This consists of fossilised tree stumps from a forest that existed here 140 million years ago, another of the wonders of the Jurassic Coast. The main footpath now traverses many miles in a roller-coaster fashion up to great heights, with great views, and down again and the prospect of another climb ahead. This makes for quite slow-going if you want to cover the miles, and can of course be exhausting, but just think of the good it's doing you!

Presently, you pass an Iron Age hillfort in the process of falling into the sea. This is Flowers Barrow and said to be haunted by the phantom legions of marching soldiers, although sightings are few and far between. The spectral legion was supposedly seen in 1678, in 1939 and several times during the 1950s and '60s. The last alleged sighting I know of was back in 1970. So don't hold your breath. If you are looking for spectacle, you may find the views of Mupe Bay, Worbarrow Bay and the striking and precipitous Gad Cliff beyond of more interest. Less well-known and photographed than Durdle Door and Lulworth Cove but equally spectacular, the Jurassic Coast is at its best here. However, if spooky things attract you, take the footpath from here to Tyneham, the 'ghost village'.

The Ghost Village of Tyneham

This small village, which once had a population of around 250 people, was evacuated by the military in 1943 to make way for tank crews practicing for the D-Day landings the following year. At the time, the villagers were told they could return to their homes when the war was over. However, in 1948, the decision was made to reserve the area for military use, although it is now open to the visiting public when not in use for Army training exercises, that is most weekends.

Here you can wander around the houses, church, streets and schoolhouse, which are said to be just as they were left in 1943. However, that is not strictly true; the village has been smartened up for visitors, the grass in the churchyard is mown and the schoolrooms made to look as if the last lesson was yesterday. In fact, the school had already closed in 1932 due to lack of pupils. Still, it makes for an interesting spectacle.

James Bond: the Dorset Connection

A notable family in Tyneham were the Bonds. This has prompted at least one guidebook author to hint at a link with the fictional spy James Bond and Bond Street in London. In Ian Fleming's penultimate Bond novel, *You Only Live Twice,* an obituary to James Bond states that his ancestry could be traced back to Dorset. More recently, in the 1999 James Bond film *The World is Not Enough,* Bond, in this case played by Pierce Brosnan, mutters the phrase used as the film title, explaining that this was a family motto. This is interesting since the Bond family, whose family seat is at Holme Priory in Dorset, does indeed have this motto. So what is the connection with the fictional James Bond?

Well, Ian Fleming attended the Durnford school near Langton Matravers for a few years from the age of seven. He would no doubt have heard of the locally

St Mary's church, Tyneham.

Old schoolroom, Tyneham.

Abandoned and now roofless building at Tyneham.

prominent Bond family, indeed he must have done so to have used the motto. Fleming first mentions this in his 1963 book *On Her Majesty's Secret Service,* and the Bond family's coat of arms appears in the 1969 film version of this story. But there is more.

In an article in *The Telegraph* dated 30 October 2008, Nick Britten revealed how the current owner of Holme Priory, William Bond, has a diary from his ancestor Denis Bond that tells of an earlier John Bond who was a spy for Sir Francis Drake, an agent for Queen and country during the reign of Elizabeth I. Apparently, it was John Bond who adopted the motto 'the world is not enough' from King Philip of Spain who had originally used it. Fleming must have been aware of this, after all a spy named J. Bond with the family motto, 'the world is not enough', is surely behind the modern-day James Bond. Strangely enough, James Bond also operates during an Elizabethan period, that of Elizabeth II. Later, at his Caribbean home Goldeneye, Fleming noticed that his ornithological book *Birds of the West Indies* was by an author named James Bond, which no doubt brought it all back to him and decided the name of his future literary creation. As for Bond Street, yes this is also connected, having been named after a member of this same family, a seventeenth-century landowner, Sir Thomas Bond.

Back on the coast, the footpath takes us up the strikingly angular Gad Cliff, past Brandy Bay and down the shale cliffs of Kimmeridge.

Kimmeridge Bay

Presently, Kimmeridge Bay comes into view. The coastline consists of shale cliffs that are constantly eroding and shedding rocks rather dangerously onto the beach. Friction can cause the shale oil to ignite as it does from time to time. When this happens, every few years, glowing, smouldering rocks tumble to the beach and geologists come from near and far to observe the phenomenon. The shale rocks have sometimes been deliberately lit to allow for scientific study. At the time of writing, there has been nothing as dramatic as the 'Lyme Volcano' of 1908 or the ignition of 'Burning Cliff' in 1826, not yet anyway.

The presence of conventional oil is also evidenced by the 'nodding donkey' oil-well pump up on the cliff top. This extracts oil from Britain's only, and Western Europe's largest, onshore oil field. However, the pump at Kimmeridge is not doing all the work; the main extraction complex is at Wytch Farm near Poole Harbour. Oil was found here in 1958 and regular extraction started the following year.

The shoreline at Kimmeridge is notable for its dark, rocky platforms that stretch out to sea when exposed at low tide, allowing observation of marine life in the clear waters of the Purbeck Marine Wildlife Reserve. The reserve was established in 1978 and was the first voluntary marine life reserve in Britain. On the eastern side of Kimmeridge Bay, near a small slipway, is the Dorset Wildlife Trust's information centre. In the information centre, there is a live seabed camera that

Clavell Tower built at Kimmeridge Bay in 1820. (*By kind permission of the Landmark Trust*)

Clavell Tower, originally intended as an observatory or perhaps just a folly. The tower is now owned by the Landmark Trust and available as holiday let accommodation. (*By kind permission of the Landmark Trust*)

can be operated by the visitor. Through this, you may see all manner of underwater life, including the black-faced blenny, a rare fish found in this locality.

The slipway was built by Sir William Clavell in the hope of turning Kimmeridge into a commercial port. He had the nearby Smedmore House built in the 1630s and was probably the first person to realise the potential of shale oil as a fuel. He used Kimmeridge shale as a fuel for boiling seawater to extract salt.

This attractive and well-sited edifice is said to have inspired P. D. James' novel *The Black Tower,* and is now owned by the Landmark Trust who specialise in unique holidays in extraordinary surroundings. In 2008, the Trust completed an operation to dismantle the tower, moving it 80 feet further inland from the cliff edge and rebuilding it. The newly refurbished tower has been converted into an unusual holiday let. The tower can accommodate two people and has an impressive 360 degree view from the living room on the top floor. A 'room with a view' indeed. Full details of this and other Landmark Trust properties can be found on their website, www.landmarktrust.org.uk.

St Aldhelm's Head

The exhausting roller-coaster nature of the coast path continues as we climb up over the slumping shale of Houns Tout Cliff then down towards the pretty cove surrounded by an undercliff of boulders and scrub and dotted with fisherman's huts known as Chapman's Pool. We then move up again onto Emmetts Hill. Presumably, this was once the haunt of a notable population of wood ants, as 'emmet' is an old name for this insect. From Emmetts Hill, we ascend St Aldhelm's Head, also known as St Alban's Head, where a strange-looking, squat building can be seen at its summit. This is St Aldhelm's chapel. So who was St Aldhelm and why has his chapel been built unconventionally, with its four corners lined up with the cardinal points? The first question is easier to answer than the second. St Aldhelm was Bishop of Sherborne in the late seventh century, who was appointed first bishop of Dorset by King Ine of Wessex. His chapel, 350 feet above the sea, is an unusual, small, square church built of Purbeck stone. As for its unusual shape and orientation, it has been suggested that the building may originally have been some kind of navigational aid, but no one is really sure. During the summer, there is a monthly service held in the chapel.

If the chapel was intended to be of assistance to passing mariners, it has a nearby modern equivalent. Just a little seaward is the National Coastwatch Institution (NCI) lookout. Staffed by trained volunteers, the NCI watch station acts as the unofficial eyes and ears of HM Coastguard in this area, following cutbacks to the number of official coastguard watch stations some years ago. A shipwreck in Cornwall, just underneath an abandoned HM Coastguard lookout, proved the need for the continued presence of visual observers at such locations, hence the formation of the NCI. The NCI now has active stations all around

Left: National Coastwatch flag on a breezy day.

Below: NCI exercise off St Aldhelm's Head with their lookout and the St Aldhelm's Chapel visible on the clifftop.

the British coastline. In recognition of the dedication of this band of trained watchkeepers in providing such an important service on a purely voluntary basis, many branches of the organisation, including St Aldhelm's, were presented with the Queen's Award for Voluntary Service in recent years. There are also NCI watch stations at other locations along the Jurassic Coast, namely Lyme Bay, Portland Bill and Swanage. These have also been awarded the aforementioned Queen's Award. At Portland Bill, the award was presented by none other than HRH the Princess Royal at a ceremony on 2 June 2012.

As of 18 February 2014, the Lyme Bay watch station, situated in the garden of the Burton Cliff Hotel at Burton Bradstock, has not functioned. It was totally destroyed by strong winds in a fierce storm. At the time of writing, alternative arrangements are being sought. Such disasters are to be expected from time to time, given the exposed nature of many of the locations of NCI posts. This is not the first setback of this kind, nor will it be the last. A few years ago, a NCI post at Folkestone in Kent had to be abandoned due to its imminent collapse over an eroding cliff edge. Undaunted, the local NCI used the top room of a nearby Martello Tower until a portacabin could be procured. This, too, found itself precariously near the cliff edge as time went on, but after a lengthy fundraising exercise a nearby Second World War bunker was converted and now provides a superb lookout in an excellent and much safer location. All the hard work paid off and the proud watchkeepers received their Queen's Award at a ceremony in October 2011. I hope the Burton Bradstock NCI are equally successful in ensuring the continuance of their vital operation.

Inland from St Aldhelm's Head, a footpath leads to the attractive village of Worth Matravers. Here you will find cottages built of Purbeck stone, a large village pond and the Square and Compasses public house, a welcome refreshment oasis before continuing the journey east.

6

To Swanage and Studland

From St Aldhelm's Head to Durlston Head marks the southern coastal boundary of the Isle of Purbeck, a stone plateau of Purbeck limestone with some Portland stone. Stone quarrying has gone on for centuries here and the caves in and on the cliffs along this coast are unmistakeable. The raw materials for such important buildings as Westminster Abbey and Canterbury Cathedral have partly come from here.

Quarrying and Wild Swimming

Stone extraction ceased in the 1950s, and many of the caves are now covered with gates or grills to both deter adventurous individuals and encourage and protect bat species that want to roost here. The rare greater horseshoe bat is resident in these caves. Below the cliffs runs a stone ledge, part of which is named Dancing Ledge. It has been surmised quite reasonably that this refers to the way that waves hitting the platform seem to dance along the ledge.

A 'quarryman's' bathing pool was dynamited into existence by Eric and Geoffrey Warner on this ledge for use by the nearby boy's preparatory school at Spyways early in the twentieth century. This is still used by some for safe bathing today, as the cliffs here otherwise plunge straight down into deep water. This little pool is where the James Bond novelist Ian Fleming learned to swim as a boy. It also formed a pleasant stop on writer, naturalist and film-maker Roger Deakin's *A Swimmer's Journey Around Britain*, published in 2000 under the title *Waterlog*. Deakin said in his book that the Dorset coast provided some of the finest sea bathing in Britain. He once tried out an ambitious 'aquatic ramble' when he swam off Studland Bay, Dancing Ledge, Kimmeridge Bay, Lulworth Cove, Stair Hole, Durdle Door and Ringstead Bay in the space of a few days. On other occasions, he swam off Portland Bill and Chesil Beach. These latter two are not recommended due to dangerous currents, and Deakin wisely warned against swimming there. However, he was particularly impressed by the pool at Dancing Ledge and its wild, idylliclocation. He described 'lying on warm grey rocks amongst ammonites the size of car tyres'. Evidently, this is a good place not only for swimming but fossil hunting too. Film-maker Derek Jarman, a former pupil of the Spywaysschool, remembered his swims here as a boy with such affection that he titled his autobiography *Dancing Ledge*.

Durlston Country Park

Heading ever eastwards brings us along the seaward side of the 280-acre Durlston Country Park, now a National Nature Reserve and set up originally by the Swanage based entrepreneur George Burt, who made his fortune from the stone industry. Along these cliffs you will find occasional little shelters containing charts and pictures of dolphins, whales and the like to help identify what you may see as you gaze out to sea. These shelters are actually intended as dolphin watch shelters, set up for the purpose of providing a suitable watch point for watching and logging dolphins and other marine life. This is all part of the Durlston Marine Project.

Within the country park itself is a visitor centre where you can listen to conversations between dolphins, porpoises, whales and other marine creatures via hydrophones placed on the seabed a quarter of a mile out to sea. A webcam also allows the visitor to watch guillemots and other seabirds on the cliff ledges.

At Anvil Point, still on the edge of Durlston Country Park, you will see the shortest lighthouse you've probably ever seen. The Anvil Point lighthouse, operated by Trinity House, is under 40 feet high and remarkably squat in shape, although it is visible enough from the sea as it stands upon 100 foot high cliffs. Since 1991, like most lighthouses today, Anvil Point has become fully automated.

Beyond Anvil Point, at Durlston Head itself, can be found the Great Globe, a huge sphere of Portland stone 10 feet in diameter and weighing 40 tons. It is marked with the earth's continents and other geographical features. This has not been carved from one solid block of stone but many sections all skilfully put together at John Mowlem's stone yard in Greenwich and placed here in 1887. The nearby Durlston Head Castle is not a military installation at all but has just been built in the semblance of a castle back in 1890 with the idea of it being a restaurant, a function that it still fulfils along with a new role as a Jurassic Coast information and exhibition centre. A new feature of the park is an astronomy centre set up and run by the Wessex Astronomical Society to take advantage of the dark skies of Durlston. This includes an observatory featuring a 14-inch Meade telescope that can be used by visitors.

Heading toward Swanage, we pass another outpost of the National Coastwatch Institution. As always, you will be welcome as a visitor unless there is an emergency or an exercise or training in progress. An annex to this watch station contains interesting and useful coastal information.

Swanage: 'Old London by the Sea' and Enid Blyton

Swanage is a small, attractive seaside resort curving around a sandy bay. It boasts safe bathing and sailing, a modest pier that reopened following restoration with help from the National Lottery Heritage Fund, an annual folk festival and all the usual seaside entertainments. The town earned its nickname 'Old London by the Sea' as it has inherited a lot of 'hand-me-down' items from the capital. For instance, the ornate

Great Globe at Durlston Head. (*Photograph by Becky Stares*)

Red Arrows over Swanage. (*By kind permission of Swanage Tourist Information Centre*)

Sunset view across Swanage Bay. (*Photograph by Robin Boultwood*)

Wellington Clock Tower once stood at London Bridge railway station until local businessman George Burt decided to bring it south, while the frontage of the town hall came from Mercers Hall in Cheapside. There are various gaslights, statues and columns all around the town that have been brought at various times from London. The Swanage Heritage Centre on the seafront has full information, including a leaflet to help you enjoy and make the most of your exploration of the town.

Between 1931 and 1965, Swanage was the favoured holiday haunt of children's author Enid Blyton. At first, she used to come on holiday and write many of her books here, but then in 1951 her second husband bought the Isle of Purbeck Golf Club, which gave them a more permanent base. Until the club was sold in 1965, Blyton would often be seen writing there. She certainly drew on her local surroundings in her stories, many of which were set in Dorset. Enid Blyton is probably most remembered for her 'Famous Five' and 'Secret Seven' series of children's adventures, along with her characters Noddy, Big Ears and Mr Plod.

Incredibly, Blyton wrote and had published some 800 books over her forty-year writing career. Although some have questioned the literary quality and political correctness of her books in recent years, she is still popular. Indeed, on 6 March 2014, she was named as the most popular children's author ever. Whether the poll respondents who voted that way were largely adults who remembered Blyton's work from their childhood, I do not know.

Another tourist attraction in Swanage is the steam train service that runs out to the romantic ruin of Corfe Castle. This is operated by the Swanage Railway Trust, which has introduced a number of steam and diesel locomotives on the line.

Swanage seafront. (*Photograph by Robin Boultwood*)

Swanage Railway Trust steam locomotive. (*Copyright Andrew P. M. Wright of Swanage Railway Trust*)

The Trust is working towards opening and operating the line as far as Wareham, where it would connect to the mainline rail network.

Corfe Castle itself is a grey, tumbled structure that sits atop a little hill in a most dramatic fashion. The grey Purbeck limestone and deserted ambience make for a sombre and striking sight. Originally built in the eleventh century, the castle was in use until the English Civil War, when it was besieged by the Parliamentarians and severely damaged in the process. Corfe Castle has remained a ruin ever since, and today is owned by the National Trust and open to visitors.

An interesting custom takes place at Corfe Castle every Shrove Tuesday. The Freemen of the Ancient Order of Purbeck Marblers hold their annual court to introduce new Freemen to the Order. After the 'official' ceremony, a game of football is played down the road to Swanage and back to Corfe Castle, with the purpose of maintaining the ancient right of way to Swanage Harbour, from where Purbeck marble was formerly shipped.

Corfe Castle with passing steam train. (*Copyright Andrew P. M. Wright of Swanage Railway Trust*)

Ballard Down, Paul Nash and Old Harry Rocks

Stretching north east of Swanage is an area of downland fronted by chalk cliffs known as Ballard Down. This is National Trust land, well-maintained chalk grassland rich in wild flowers and butterflies, including the beautiful but rare Chalkhill Blue and Adonis blue. Herring gulls and Great Black-backed Gulls (the largest British gull) nest on the cliffs.

Ballard Down is the subject of at least two paintings by artist Paul Nash, completed during his 'Swanage Period' between 1933 and 1936. *Event on the Downs* is a semi-surrealist view across the downs to Ballard Cliff from Whitecliff Farm, just outside Swanage where Nash stayed in 1933 and 1934/35, while *Voyage of the Fungus* also has a landscape background that includes Ballard Cliff. Nash painted other Dorset scenes too, both in surreal and realistic styles, at Swanage, West Lulworth, Kimmeridge and the countryside near Worth Matravers. He was also commissioned by John Betjeman to write the *Dorset Shell Guide,* which was first published in 1936. This very individual guide was illustrated with many of Nash's own watercolours and photographs. In 1935, Nash moved into Swanage itself and lived at No. 2 The Parade, but moved away the following year.

The cliffs themselves start as a small, vegetated undercliff just outside Swanage and gradually become the sheer, white cliff faces of Ballard Cliff and beyond. Towards Studland, chalk stacks known as the Pinnacles stand sentinel just offshore. To me, these are oddly reminiscent of the stubby chalk cliffs of Kingsgate Bay and Botany Bay on the North Foreland between Broadstairs and Margate in my home county of Kent, which are similar and also prone to forming stacks and archways.

Following these are a line of stacks known as Old Harry Rocks. The outer two of these, known as 'Old Harry and his wife', once formed a chalk arch. This was destroyed by the sea in 1896 leaving just the two stacks. The reference to 'Harry' probably refers to King Henry VIII, who had a castle built here in 1538 that has also since been washed away by the sea. These cliffs have their counterparts across the sea at the Needles on the Isle of Wight, to which they were connected before the sea level rose at the end of the last Ice Age. The Old Harry Rocks mark the official end of the Jurassic Coast (if travelling east), and beyond lies Studland, where the chalk cliffs diminish to nothing and are replaced by sand dunes.

Old Harry Rocks from the sea. (*Photograph by Jackie Lane*)

Studland and Beyond

Today, this area is owned and managed by the National Trust and the beaches are frequented by large numbers of holidaymakers and popular with nudists. The dunes apparently resemble the beaches of Normandy and so were used for rehearsing the D-Day landings in 1944. These dunes and the grassland behind are said to support more wild flowers per acre than anywhere else in Britain. They are also home to all the classic British reptile species, namely the common lizard, adder, grass snake and slow worm, along with the rare sand lizard and smooth snake. These are said to be the 'six' British reptile species. However, if we include the Channel Islands, there is a seventh, the green lizard, the largest British lizard. Also, at various places in southern England along the coast and inland, there are substantial established colonies of wall lizards, making at least eight species (and probably more, as I don't think all these wall lizards are of the same species). Naturalists will argue that wall lizards are not native, but of course they are now. After all, there were no reptiles in Britain at all during the last Ice Age, so they have all arrived in the last few thousand years by one means or another. But lets not be pedantic, six species in one area, including two that are rare, is pretty good, and proves the value of this site for the preservation of British reptiles.

Swanage Sea Fret. (*Photograph by kind permission of Jackie Lane*)

Beyond Studland lies the vast expanse of Poole Harbour, one of the largest natural harbours in the world with a 100 miles of coastline around its perimeter, much of it designated as nature reserves. Within the expansive confines of the harbour can be seen Brownsea Island, a 500-acre nature reserve with a 3-mile coastline owned by the National Trust, where red squirrels still survive. Brownsea Island is also well-known as the place where the Scout movement began. It contains one of Britain's largest heronries and receives an astonishing 100,000 visitors per year. Across the harbour, we can see the prestigious seaside homes of Sandbanks and the large and prosperous resort of Bournemouth. To the south east is the Isle of Wight, which of course has its own geologically and historically interesting coastline. All this, however, lies beyond the scope of this book and makes another interesting story, but one for another time.

Swanage Bay at sunset. (*Photograph by Robin Boultwood*)

Sunrise at Old Harry Rocks. (*Photograph by Becky Stares. Copyright Becky Stares*)

Above: Looking towards the Pinnacles and Swanage. (*Photograph by Becky Stares. Copyright Becky Stares*)

*Overleaf:*Corfe Castle with steam locomotive.

Chronology

Some Notable Dates in the History of the Jurassic Coast

From 250 Million Years BP (Before the Present)
The Triassic period. The earliest reptiles and dinosaurs live and red sandstone is laid down, now visible in the cliffs of East Devon.

From 210 Million Years BP
The Jurassic period, which gives its name to this coastline. Dinosaurs rule the earth. Shales, limestones and sandstones are laid down and, now visible at many locations along the Jurassic coast.

From 145 Million Years BP
The Cretaceous period and the peak of dinosaur development. The first birds and flowering plants appear. Many types of ammonite flourish in the warm seas. This period ends abruptly with the impact of an asteroid in what is now the Gulf of Mexico. Dinosaurs and many other life forms were wiped out and mammals get their chance. Chalk was laid down during this period.

8000–4000 BC
The Mesolithic period. Early settlers leave traces on the Isle of Portland from now on.

6500 BC
The English Channel is formed as rising sea levels cut Britain off from the European continent. Erosion by the sea creates the cliffs along the Channel coast that we see today.

c. 700 BC
The Iron Age. Stoneworking may have started on Isle of Portland. Swords from this period have been found in Weymouth Bay.

AD 43–c. 410
The Roman period. Seaton in East Devon is an important port. Stone quarrying starts in earnest at Beer in Devon and on the Isle of Portland. Dorchester, a major Roman town, lies just inland.

AD 390–420
Romano-British temple in use at Jordan Hill near Osmington.

Early Eighth Century
St Aldhelm, Bishop of Sherborne appointed Bishop of Dorset by King Ine of Wessex. A chapel dedicated to him stands at St Aldhelm's Head.

831
15,000 Vikings land at Charmouth and may be responsible for the martyrdom of St Wite, to whom the church at Whitchurch Canonicorum near Morcombelake is dedicated some fifty years later by King Alfred. The name 'Wite' is latinised as 'Candida' in the thirteenth century.

Eleventh Century
Benedictine monastery established at Abbotsbury.

Late Twelfth Century
Rufus Castle on Isle of Portland built during the reign of William II, nicknamed 'Rufus' because of his red hair.

1211
Earliest record of rope making at Bridport.

1284
Lyme Regis is granted its 'Regis' title when the town receives a royal charter from Edward I.

1393
The first record of a herd of swans at Abbotsbury.

1538–1540
Henry VIII has a line of defensive castles built around Britain, including Portland Castle and Sandsfoot Castle in Weymouth.

1554
Sir Walter Raleigh born near Budleigh Salterton.

1645/46
The siege of Corfe Castle by Parliamentarians during the English Civil War takes place. The castle is severely damaged in the siege and remains a ruin to the present day.

1666
The Great Fire of London. After this, architect Sir Christopher Wren, a one time MP for Weymouth designs a range of new buildings for London. Many of them, such as St Paul's Cathedral, are subsequently built using Portland stone, as are other public buildings such as Buckingham Palace and, more recently, the United Nations Headquarters in New York.

1685

The Duke of Monmouth lands near Lyme Regis and leads an unsuccessful attempt to dethrone James II.

1789–1805

George III uses Weymouth (previously a small fishing village) as his summer retreat and thereafter the place becomes a popular holiday resort. In Weymouth, the King becomes the first monarch to use a bathing machine.

1790

Substantial landslip at Hooken Cliffs between Branscombe and Beer creates a dramatic landscape of undercliff and chalk pinnacles.

1803–04

Jane Austen visits Lyme Regis. She sets part of her novel *Persuasion* in the town.

1811

Twelve-year-old Mary Anning discovers a fossil icthyosaurus (a prehistoric marine reptile) as well as a plesiosaurus and a pterosaur (a flying reptile) below Black Ven near Lyme Regis. This starts the craze for fossil hunting that continues to this day.

1819

The Duke and Duchess of Kent, escaping their creditors, move to Sidmouth, bringing their daughter the future Queen Victoria with them.

1820

Clavell Tower is built at Kimmeridge Bay, possibly as an observatory or maybe just a folly. In 2008, the tower is moved inland to prevent its collapse over the cliff edge and has been converted into a holiday let by The Landmark Trust who specialise in unusual holiday accommodation.

1824

Storm surge breaches Chesil Beach and sweeps away the village of East Fleet. The village inspired the J. Meade-Falkner novel *Moonfleet*.

1826

Spontaneous combustion of oil-rich shale in cliffs near Ringstead Bay takes place. The cliffs smoulder for years and subsequently become known as 'Burning Cliff', a name they have retained to this day.

25 December 1839

A massive landslip occurs at Bindon in the undercliff between Axmouth and Lyme Regis. A sensation in its time, it even inspires a piece of music, *The Landslip Quadrille*.

3 February 1840
Another major landslip occurs at Whitlands in the Axmouth to Lyme Regis undercliff.

1849–1872
Construction of Portland Harbour, one of the world's largest artificial harbours.

1868–1897
Thomas Hardy writes a number of novels set in Dorset and along its coast. Actual locations are thinly disguised by slightly altering names.

1886
Durlston Country Park is developed by entrepreneur George Burt upon his retirement. The Great Globe is installed in 1887 and Durlston Head Castle built in 1890.

1898
J. Meade-Falkner's novel *Moonfleet* set on the Dorset coast is published.

1908
Following a landslip at Black Ven, spontaneous combustion occurs in a conical formation that becomes known as the 'Lyme volcano'.

1929–1934
Artist John Piper visits Portland and Chesil Beach and produces a number of works featuring the area.

1931–1965
Enid Blyton, often in Swanage, writes many of her children's adventure books here.

1933–36
Artist Paul Nash is in Swanage, firstly at White Cliff Farm and later in the town itself. He uses the locality in some of his paintings during what he referred to as his 'Swanage period'.

1943
The village of Tyneham is requisitioned by the military for training troops in preparation for the D-Day landings. Villagers are evacuated from their homes with a promise they can return after the War is over. The promise has never been honoured and the village is today still in the hands of the Army, although it is open to the public at weekends and public holidays.

1944
Training for the D-Day landings takes place among the dunes at Studland as well as the ranges near Lulworth. The Expeditionary Force leaves for France from Portland Harbour.

1949–1984
The National Trust acquires areas of land in stages at Ringstead Bay and White Nothe.

1955
Britain's first national nature reserve is established at the Axmouth to Lyme Regis undercliff.

1957–1958
A huge landslip occurs at Black Ven, near Lyme Regis. At the time, this was said to be the largest mudslide in Europe.

1958
Oil is discovered beneath the Isle of Purbeck. A major facility is established at Wytch Farm near Poole Harbour and drilling and pumping operations are established at other locations too, including at Kimmeridge Bay. Commercial production starts the following year. At the time, this was said to be the largest onshore oil field in Western Europe.

1961–1994
The National Trust gradually acquires areas of land around Golden Cap.

1967
The first film version of Thomas Hardy's *Far from the Madding Crowd* is released.

1969
John Fowles' novel *The French Lieutenant's Woman*, set largely around Lyme Regis is published. The book is turned into a film in 1980.

1978
Dorset Wildlife Trust's Purbeck Marine Wildlife Reserve is set up at Kimmeridge Bay.

1982
Tout Quarry on the Isle of Portland is closed commercially but reopened the following year by the Portland Sculpture and Quarry Trust as an active sculpture park, and subsequently as a nature reserve run by the Dorset Wildlife Trust.

2001
The Jurassic Coast is established as a UNESCO World Heritage Site. Inuguaration takes place the following year with the unveiling of Michael Fairfax's 'geoneedle' sculpture by HRH The Prince of Wales.

18 January 2007
MSC *Napoli*, a large container ship, is wrecked off Branscombe leading to large numbers of containers washing up on the beach and spilling their contents. Many people came from far and wide to take away the stranded goods.

2007
Ian McEwan's best-selling novel *On Chesil Beach* is published.

2012
The Olympic and Paralympic sailing events take place in Weymouth Bay and Portland Harbour.

2 June 2012
The Queen's Award for Voluntary Service is presented to the National Coastwatch Institution (NCI) at Portland Bill. Other NCI branches along the Jurassic Coast and around the British coastline generally have also received the award in recent years.

2012/2013
Holloway by Robert Macfarlane, Stanley Donwood and Dan Richards, detailing the experiences of the three men while exploring south Dorset's ancient sunken footpaths, is published.

October 2013–February 2014
A season of severe storms affects the south west coast as elsewhere. Landslips uncover many fossils, including a complete icthyosaurus at Black Ven. There is much coastal erosion and a notable rock stack on the Portland coast, known as Pom Pom Rock, is totally destroyed. The narrow neck of land that connects the Isle of Portland to the 'mainland' is swamped by the sea for the first time since 1980, and customers have to be led to safety from the Cove House Inn. Heavy rain causes landslips that result in many paths, including that through the Axmouth to Lyme Regis undercliff, being closed. On 18 February, the National Coastwatch Institution Watch Station above Burton Beach is totally destroyed by high winds.

Bibliography

Townshend, C. *Abbotsbury Swannery and the Fleet* (Abbotsbury Swannery and Jarrold Publishing, 1987).

Barber, Donald, 'Invasion by Washing Water' (article in *Perspective* magazine, vol. 5, No. 4, 1968).

Coram, R. BA Hons (Oxon), *Where to Find Fossils in Southern England: A Guide to Classic Localities* (British Fossils, 1989).

Campbell, Donald, *Exploring the Undercliffs: The Axmouth to Lyme Regis National Nature Reserve, A 50th Anniversary Guide* (Coastal Publishing, 2006).

Damon, Robert F. G. S., *The Geology of Weymouth and the Isle of Portland (with notes on the Natural History of the Coast and Neighbourhood)* (Edward Stanford, 1880).

Deakin, Roger, *Waterlog: A Swimmer's Journey Through Britain* (Vintage Books, 2000).

Eates, Margot, *Paul Nash: The Master of the Image 1889–1946* (John Murray Publishers Ltd, 1973).

Fowles, John, *The French Lieutenant's Woman* (Jonathan Cape, 1969).

Hancock, Mathew and Amanda Tomlin, *The Rough Guide to Dorset, Hampshire and the Isle of Wight* (Rough Guides, 2010).

Hardy, Thomas, *Far from the Madding Crowd* (Penguin Classics, 2000, originally published in 1874).

Hardy, Thomas, *The Trumpet Major* (Penguin Classics, 2000, originally published in 1880).

Hardy, Thomas, *The Well Beloved* (Penguin Classics, 2000, originally published in 1897).

Hollands, Ray, *Along the Dorset Coast* (The History Press, 2010).

Ingrams, Richard and John Piper, *Piper's Places: John Piper in England and Wales* (Chatto & Windus, The Hogarth Press, 1983).

Legg, Rodney, *Dorset's Jurassic Coast* (PIXZ Books, 2008).

Macfarlane, Robert, Stanley Donwood and Dan Richards, *Holloway* (Faber & Faber, 2013).

Marren, Peter and Richard Mabey, *Bugs Britannica* (Chatto & Windus, 2010).

McEwan, Ian, *On Chesil Beach* (Jonathan Cape, 2007).

McCloy, Andrew, *Britain's Best Coastal Walks* (UK: New Holland Publishers Ltd, 2011).

Meade-Falkner, John, *Moonfleet* (Edward Arnold, 1898).

Nash, Paul, *Dorset Shell Guide* (Architectural Press, 1936).

Smith, Roly, *World Heritage Sites of Britain* (AA Publishing, 2010).

Soper, Tony, *A Natural History Guide to the Coast* (Peerage Books, 1984).

Start, Daniel, *Wild Swimming Coast – Explore the Secret Coves and Wild Beaches of Britain* (Punk Publishing Ltd, 2009).

Syer, Canon G. V., *Church of St Candida and Holy Cross, Whitchurch Canonicorum, Dorset, The Cathedral of the Vale, Church Guide and History* (Church of St Candida and Holy Cross, 1981).

Tarr, Roland, *South West Coast Path, Exmouth to Poole* (Aurum Press, 2009).

Paul Nash (Tate Gallery, 1975).

Treves, Sir Frederick, *The Highways and Byways of Dorset* (Macmillan, 1914).

Various, *Folklore, Myths and Legends of Britain* (Reader's Digest, 1973).

Various, *The Illustrated Guide to Britain's Coast* (Drive Publications Ltd, 1987).

Various, *Secret Britain* (Automobile Association, 1986).

Walters, Christine, *Who was St Wite? The Saint of Whitchurch Canonicorum* (Christine Walters, 1980).

Wilkins, George A., *Donald Barber* (obituary in *The Independent,* 28 August 2000).

Wickramasinghe J. T. and N. T. Wickramasinghe 'On the Possibility of the Transfer of Microbiota from Venus to Earth' (article published online, 2008).

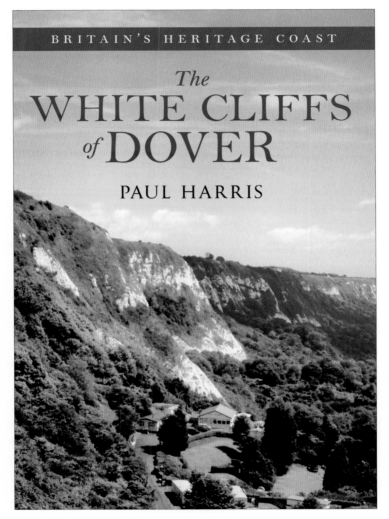